Musty-Crusty
Animals

Musty-Crusty Animals ABC

Lola M. Schaefer

Heinemann Library
Chicago, Illinois

Customer Service 888-454-2279
Visit our website at www.heinemannlibrary.com

Designed by Sue Emerson/Heinemann Library and Ginkgo Creative, Inc.
Printed and bound in the U.S.A. by Lake Book

06 05 04 03 02
10 9 8 7 6 5 4 3 2 1

Library of Congress Cataloging-in-Publication Data
Schaefer, Lola M., 1950-
    Musty-crusty animals ABC / Lola Schaefer.
        p. cm. — (Musty-crusty animals)
Includes index.
Summary: An alphabet book providing factual information on sea horses and various crustaceans.
    ISBN 1-58810-518-0 (lib. bdg.)   ISBN 1-58810-727-2 (pbk. bdg.)
    1. Crustacea—Juvenile literature. 2. Sea horses—Juvenile literature. 3. Limulus polyphemus—Juvenile literature.
    4. English language—Alphabet—Juvenile literature. [1. Crustaceans. 2. Sea horses. 3. Alphabet.] I. Title.
    QL437.2 .S32 2002
    595.3—dc21
    [[
                                        2001003287

Acknowledgments
The author and publishers are grateful to the following for permission to reproduce copyright material:
p. 3 Rudie Kuiter/Seapics.com; p. 4 Doug Perrine/Seapics.com; p. 5 Gary Meszaros/Bruce Coleman Inc.; p. 6 Jane Burton/Bruce Coleman Inc.; p. 7 Jeff Rotman Photography; p. 8 D. Lyons/Bruce Coleman Inc.; p. 9 Dwight Kuhn; p. 10 John G. Shedd Aquarium/Visuals Unlimited; p. 11 Bryan Hitchcock/National Audubon Society/Photo Researchers, Inc.; p. 12 Ronald Sefton/Bruce Coleman Inc.; p. 13 Jonathan Bird/ORG; p. 14 Rudie Kuiter/Seapics.com; p. 15 Greg Ochocki/ Seapics.com; p. 16 Patrice Ceisel/Visuals Unlimited; p. 17 Link/Visuals Unlimited; p. 18 Philip Gould/Corbis; p. 19 Doug Perrine/Seapics.com; p. 20 Grace Davies Photography; p. 21 Triarch/Visuals Unlimited; p. 22 Kazunari Kawashima

Cover photographs courtesy of (L–R): Grace Davies Photography; Dwight Kuhn; Patrice Ceisel/Visuals Unlimited

Every effort has been made to contact copyright holders of any material reproduced in this book. Any omissions will be rectified in subsequent printings if notice is given to the publisher.

Special thanks to our advisory panel for their help in the preparation of this book:
Eileen Day, Preschool Teacher
Chicago, IL

Paula Fischer, K–1 Teacher
Indianapolis, IN

Sandra Gilbert,
Library Media Specialist
Houston, TX

Angela Leeper,
Educational Consultant
North Carolina Department
of Public Instruction
Raleigh, NC

Pam McDonald, Reading Teacher
Winter Springs, FL

Melinda Murphy,
Library Media Specialist
Houston, TX

Helen Rosenberg, MLS
Chicago, IL

Anna Marie Varakin,
Reading Instructor
Western Maryland College

Special thanks to Dr. Randy Kochevar of the Monterey Bay Aquarium for his help in the preparation of this book.

Some words are shown in bold, **like this.**
You can find them in the picture glossary on page 23.

A a   Adult

pouch

Adult male sea horses carry eggs in their **pouches**.

B b Barnacle

barnacles

Barnacles live on rocks, wood, and some animals.

C c Crayfish
D d Daytime

Crayfish hide in the daytime.

They hide under rocks and logs.

E e   Eight
F f   Food

Lobsters have eight walking legs.

They crawl to hunt for food.

G g Gills
H h Horseshoe Crab

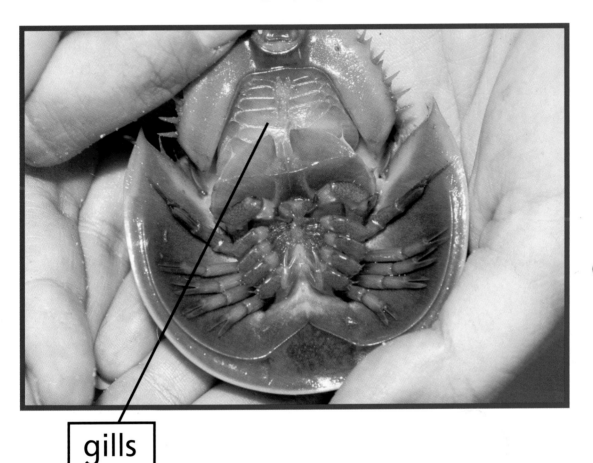

gills

Gills help horseshoe crabs breathe underwater.

Ii   Instar

instar

Little crayfish are called **instars**.

Jj   Jointed Legs

jointed leg

Hermit crabs have **jointed legs**.

The legs can move many ways.

K k   Knob

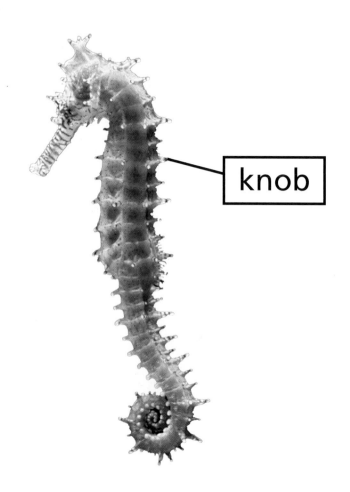

knob

Many sea horses have **knobs**
on their bodies.

L l   Lobster
M m   Molt

Lobsters molt.

They can grow new shells.

N n　Nip

Hermit crabs have sharp claws.

They can nip you.

O o   Octopus

An octopus will crush and eat
hermit crabs.

P p   Pop! Pouch

| pouch | young sea horse |

Pop! Young sea horses pop out of the **pouch** one at a time.

Q q   Quick
R r   Rocks

Lobsters curl their tails for a quick getaway into the rocks.

S s   Snout

snout

Sea horses have long **snouts.**

They suck food out of the water.

T t   Tide

Sea water covers barnacles at high tide.

The tides bring food to the barnacles.

U u   Underside

eggs

Crayfish lay eggs on their undersides.

The eggs look like small blueberries.

V v   Vision

Sea horses have good vision.

They can see well.

W w Warm Waters
X x Exoskeleton

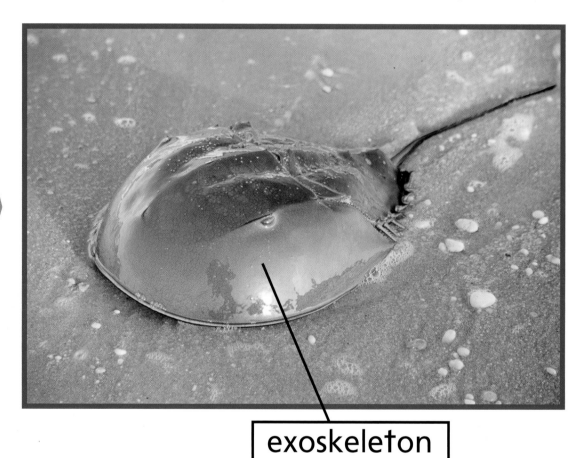

exoskeleton

Horseshoe crabs live in warm waters near land.

They have large **exoskeletons**.

Yy   Young

Young barnacles come out of their eggs.

They swim away.

Z z   Zoea

A very young hermit crab is called a **zoea**.

Picture Glossary

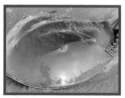 **exoskeleton**
(EX-oh-SKELL-uh-tuhn)
page 20

 **gills**
page 7

 **instar**
page 8

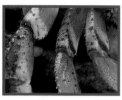 **jointed legs**
page 9

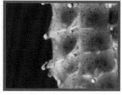 **knob**
page 10

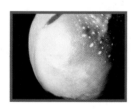 **pouch**
pages 3, 14

 **snout**
page 16

 **zoea**
(zo-EE-uh)
page 22

Note to Parents and Teachers

Using this book, children can practice alphabetic skills while learning interesting facts about musty-crusty animals. Together, read *Musty-Crusty Animals ABC.* Say the names of the letters aloud, then say the target word, exaggerating the beginning of the word. For example, "/b/: Bbbb-arnakul." Can the child think of any other words that begin with the /b/ sound? (Although the letter x is not at the beginning of the word "exoskeleton," the /ks/ sound of the letter x is still prominent.) Try to sing the "ABC song," substituting the musty-crusty alphabet words for the letters a, b, c, and so on.

! CAUTION: Remind children that it is not a good idea to handle wild animals. Children should wash their hands with soap and water after they touch any animal.

Index